Friedrich Krinzeßa

Fast construction and evaluation of interpolatory periodic spline curves

GRIN Verlag

Bibliografische Information der Deutschen Nationalbibliothek:

Die Deutsche Bibliothek verzeichnet diese Publikation in der Deutschen National-
bibliografie; detaillierte bibliografische Daten sind im Internet über http://dnb.d-
nb.de/ abrufbar.

Imprint:

Copyright © 2006 GRIN Verlag GmbH
Druck und Bindung: Books on Demand GmbH, Norderstedt Germany
ISBN: 978-3-638-91976-0

This book at GRIN:

http://www.grin.com/en/e-book/87659/fast-construction-and-evaluation-of-interpo-
latory-periodic-spline-curves

GRIN - Your knowledge has value

Der GRIN Verlag publiziert seit 1998 wissenschaftliche Arbeiten von Studenten, Hochschullehrern und anderen Akademikern als eBook und gedrucktes Buch. Die Verlagswebsite www.grin.com ist die ideale Plattform zur Veröffentlichung von Hausarbeiten, Abschlussarbeiten, wissenschaftlichen Aufsätzen, Dissertationen und Fachbüchern.

Fast Construction and Evaluation of Interpolatory Periodic Spline Curves

Friedrich Krinzeßa

August 2006

CONTENTS

0   Introduction..3

1   Interpolatory periodic cubic B-spline curves in Bernstein Bézier Form..........4

2   Interpolatory periodic cubic B-spline curves in de Boor Form...................16

3   Numerical calculation..18

4   References...22

0. INTRODUCTION

In this publication, an explicit representation of formulas for periodic cubic spline interpolation by curves in $\mathbb{R}^2$ and $\mathbb{R}^3$ is given for the classical case where data points and nodal points coincide. The solution is formed using Bézier points and basic splines. Furthermore, interpolation with equidistant parameters is discussed. Of course, the achieved results can be used for numerical calculation.

In the following, the result for interpolation using basic splines will be formulated a little more exactly. Let s be a $(\mathbf{x}_n - \mathbf{x}_0)$-periodic cubic B-spline curve, which interpolates in given data points $\mathbf{x}_0$, $\mathbf{x}_1$, ..., $\mathbf{x}_{n-1}$, $\mathbf{x}_n = \mathbf{x}_0 + \mathbf{p}$. Then the interpolating periodic cubic spline s in $[t_0 ; t_n]$ can be expressed as

$$s = \sum_{l=-2}^{n} \mathbf{d}_l N_{l,3}(t)$$

with the control points $\mathbf{d}_l$ and the functions $N_{l,3}(t)$ of de Boor, which are defined by recursion.

If

$$t_{i+1} - t_i = \Delta_i$$

with the knots

$$\ldots < t_{-2} < t_{-1} < t_0 < t_1 < t_2 < \ldots ,$$

one can write the control points $\mathbf{d}_l$ as

$$\mathbf{d}_l = \mathbf{x}_l + \frac{1}{v_n} \sum_{r=0}^{n-1} d_{r,l,n} \Delta \mathbf{x}_{l+r},$$

whereby the $d_{r,l,n}$ and v_n are functions of only the Δ_i. More precisely, the $d_{r,l,n}$ are linear combinations of the elements of the matrix products $G(\Delta_{l+r-1}) \cdot G(\Delta_{l+r-2}) \cdot \ldots \cdot G(\Delta_{l+1})$ and $G(\Delta_{l+n-1}) \cdot G(\Delta_{l+n-2}) \cdot \ldots \cdot G(\Delta_{l+r+1})$ and v_n is a linear combination of the diagonal elements of the matrix product $G(\Delta_{n-1}) \cdot G(\Delta_{n-2}) \cdot \ldots \cdot G(\Delta_0)$. With this

$$G(\Delta_r) = \begin{pmatrix} -2 & -\dfrac{\Delta_r}{2} \\ -\dfrac{6}{\Delta_r} & -2 \end{pmatrix} \quad r = 0, 1, 2, \ldots, n-1.$$

Similar formulas can be used when expressing this with Bézier points.

Explicit expressions can be obtained from the formulas we have just described.

As $d_{r,l,n}$ and v_n, represented by means of matrix products, take on considerable values with increasing n, the above described formulas can be applied only until $n \approx 500$ for numerical purposes. This can be remedied, however, by skillfully canceling $\dfrac{d_{r,l,n}}{v_n}$. By inserting known iterations, the calculation time can be reduced, which, for $n = 1000$, reduces the time from about 4 s to about 0.4 s.

If the parameters are equidistant, the series of values $\dfrac{d_{r,l,n}}{v_n}$ ($r = 0, 1, 2, \ldots, n-1$) do not differ for different l and fixed n. Moreover, they do not depend on Δ_i, but are thus fixed numbers. This means if the parameters are equidistant and the number n is defined from the beginning, the same fixed series of numbers $\dfrac{d_{r,l,n}}{v_n}$ $(r = 0, 1, 2, ..., n-1)$ can always be used, which will shorten the calculating time enormously.

1. INTERPOLATORY PERIODIC CUBIC B-SPLINE CURVES IN BERNSTEIN BÉZIER FORM

In this paper, matrices (cf. Zeilfelder [2, (2.7)]) of the following form and their products will play a fundamental role.

DEFINITION 1.

$$G(a) = \begin{pmatrix} -2 & -\dfrac{a}{2} \\ -\dfrac{6}{a} & -2 \end{pmatrix} \; ; \; a > 0 \; .$$

LEMMA 1. *The inverse of the matrix $G(a)$ is*

$$G^{-1}(a) = \begin{pmatrix} -2 & \dfrac{a}{2} \\ \dfrac{6}{a} & -2 \end{pmatrix} \; ; \; a > 0 \; .$$

Proof.

$$G(a) \cdot G^{-1}(a) = I \; ; \; a > 0 \; . \quad \blacksquare$$

DEFINITION 2. *If $M = \{..., a_{-2}, a_{-1}, a_0, a_1, a_2, ...\}$ is an ordered set of real positive numbers, the following is true:*

(i) $\begin{pmatrix} g_{a_i}^{a_j}(1,1) & g_{a_i}^{a_j}(1,2) \\ g_{a_i}^{a_j}(2,1) & g_{a_i}^{a_j}(2,2) \end{pmatrix} = G^{-1}(a_{j+1}) \cdot G^{-1}(a_{j+2}) \cdot ... \cdot G^{-1}(a_{i-1}) \; for \; i-j > 1 \; ,$

(ii) $\begin{pmatrix} g_{a_i}^{a_j}(1,1) & g_{a_i}^{a_j}(1,2) \\ g_{a_i}^{a_j}(2,1) & g_{a_i}^{a_j}(2,2) \end{pmatrix} = I \; for \; i-j = 1 \; ,$

(iii) $\begin{pmatrix} g_{a_i}^{a_j}(1,1) & g_{a_i}^{a_j}(1,2) \\ g_{a_i}^{a_j}(2,1) & g_{a_i}^{a_j}(2,2) \end{pmatrix} = G(a_j) \cdot G(a_{j-1}) \cdot ... \cdot G(a_i) \; for \; i-j < 1 \; .$

These products of matrices can be determined explicitly. To do this, the following definition can be used.

DEFINITION 3. *If $M = \{a_1, a_2, ..., a_n\}$ is an ordered set of real positive numbers, we define for $2 \le k \le n$ (k being an even integer)*

$$K_{a_n}^{a_1}(k) = \sum_{j_1=1}^{n} \sum_{j_2=j_1+1}^{n} ... \sum_{j_k=j_{k-1}+1}^{n} \dfrac{a_{j_1} a_{j_3} ... a_{j_{k-1}}}{a_{j_2} a_{j_4} ... a_{j_k}} \; ; \; K_{a_1}^{a_n}(k) = \sum_{j_1=1}^{n} \sum_{j_2=j_1+1}^{n} ... \sum_{j_k=j_{k-1}+1}^{n} \dfrac{a_{j_2} a_{j_4} ... a_{j_k}}{a_{j_1} a_{j_3} ... a_{j_{k-1}}} \; ,$$

for $k = 0$

$$K_{a_n}^{a_1}(k) = K_{a_1}^{a_n}(k) = 1$$

and for $1 \le k \le n$ *(k being an odd integer)*

$$K^{a_1;a_n}(k) = \sum_{j_1=1}^{n} \sum_{j_2=j_1+1}^{n} \cdots \sum_{j_k=j_{k-1}+1}^{n} \frac{a_{j_1} a_{j_3} \ldots a_{j_{k-2}} a_{j_k}}{a_{j_2} a_{j_4} \ldots a_{j_{k-1}}} \quad ; \quad K_{a_1;a_n}(k) = \sum_{j_1=1}^{n} \sum_{j_2=j_1+1}^{n} \cdots \sum_{j_k=j_{k-1}+1}^{n} \frac{a_{j_2} a_{j_4} \ldots a_{j_{k-1}}}{a_{j_1} a_{j_3} \ldots a_{j_{k-2}} a_{j_k}}.$$

This means: Maintaining the previous order, all combinations from M with k elements are written as fractions in such a way, that the elements of the combinations are written alternately into the nominator and into the denominator as a factor. Here, the first element of the k elements is written into the nominator if a_1 is on top, and into the denominator, if a_1 is on bottom. Then the last element of each combination must be where a_n is. After this, the fractions will be added.

The following table is helpful for the computerized evaluation of $K_{a_j}^{a_i}(k)$:

	a_i	a_{i+1}	$\cdots$	a_{j-k}	a_{j-k+1}	$\cdots$	$\cdots$	a_j	
.	1	1	$\cdots$	1					
:	0	$K^{a_i,a_i}(1)$	$K^{a_i,a_{i+1}}(1)$	$\cdots$	$K^{a_i,a_{j-k+1}}(1)$				
$\cdots$		$\cdots$	$\cdots$	$\cdots$	$\cdots$	$\cdots$	$\cdots$		
:			0	$K^{a_i,a_{i+k-2}}(k-1)$	$K^{a_i,a_{i+k-1}}(k-1)$	$\cdots$	$\cdots$	$K^{a_i,a_{j-1}}(k-1)$	
.				0	$K_{a_{i+k-1}}^{a_i}(k)$	$K_{a_{i+k}}^{a_i}(k)$	$\cdots$	$K_{a_{j-1}}^{a_i}(k)$	$\boxed{K_{a_j}^{a_i}(k)}$

with $K^{a_i,a_{j-1}}(1) = \ldots = K^{a_i,a_{i+k-3}}(k-1) = K_{a_{i+k-2}}^{a_i}(k) = 0$,

$K_{a_{i-1}}^{a_i}(0) = K_{a_i}^{a_i}(0) = \ldots = K_{a_{j-k}}^{a_i}(0) = 1$,

$K_{a_\sigma}^{a_\rho}(\tau) = K_{a_{\sigma-1}}^{a_\rho}(\tau) + a_\sigma^{\left((-1)^{\tau-1}\right)} \cdot K^{a_\rho,a_{\sigma-1}}(\tau-1) \quad \tau = 2,4,6,\ldots$ and

$K^{a_\rho,a_\sigma}(\tau) = K^{a_\rho,a_{\sigma-1}}(\tau) + a_\sigma^{\left((-1)^{\tau-1}\right)} \cdot K_{a_{\sigma-1}}^{a_\rho}(\tau-1) \quad \tau = 1,3,5,\ldots$.

$K^{a_i,a_j}(k) \quad k = 1,3,5,\ldots$ can be evaluated in a similar table.

Helpful for the computerized evaluation of the $K_{a_i}^{a_j}(k)$ will be the following table

	a_i	a_{i+1}	$\cdots$	a_{j-k}	a_{j-k+1}	$\cdots$	$\cdots$	a_j	
:	1	1	$\cdots$	1					
.	0	$K_{a_i,a_i}(1)$	$K_{a_i,a_{i+1}}(1)$	$\cdots$	$K_{a_i,a_{j-k+1}}(1)$				
$\cdots$		$\cdots$	$\cdots$	$\cdots$	$\cdots$	$\cdots$	$\cdots$		
.			0	$K_{a_i,a_{i+k-2}}(k-1)$	$K_{a_i,a_{i+k-1}}(k-1)$	$\cdots$	$\cdots$	$K_{a_i,a_{j-1}}(k-1)$	
:				0	$K_{a_i}^{a_{i+k-1}}(k)$	$K_{a_i}^{a_{i+k}}(k)$	$\cdots$	$K_{a_i}^{a_{j-1}}(k)$	$\boxed{K_{a_i}^{a_j}(k)}$

with $K_{a_i,a_i}(1) = \ldots = K_{a_i,a_{i+k-3}}(k-1) = K_{a_i}^{a_{i+k-2}}(k) = 0$,

$K_{a_i}^{a_{i-1}}(0) = K_{a_i}^{a_i}(0) = \ldots = K_{a_i}^{a_{j-k}}(0) = 1$,

$K_{a_i}^{a_\sigma}(\tau) = K_{a_i}^{a_{\sigma-1}}(\tau) + a_\sigma^{\left((-1)^{\tau}\right)} \cdot K_{a_\rho,a_{\sigma-1}}(\tau-1) \quad \tau = 2,4,6,\ldots$ and

$K_{a_\rho,a_\sigma}(\tau) = K_{a_\rho,a_{\sigma-1}}(\tau) + a_\sigma^{\left((-1)^{\tau}\right)} \cdot K_{a_i}^{a_{\sigma-1}}(\tau-1) \quad \tau = 1,3,5,\ldots$.

$K_{a_i,a_j}(k) \quad k = 1,3,5,\ldots$ can be evaluated in a similar table.

COROLLARY 1. *If all a_k are equal to a in DEFINITION 3, for $k \geq 2$ (k being an even integer)*

$$K_{a_n}^{a_1}(k) = K_{a_1}^{a_n}(k) = \binom{n}{k},$$

for $k = 0$

$$K_{a_n}^{a_1}(k) = K_{a_1}^{a_n}(k) = 1$$

and for $k \geq 1$ (k being an odd integer)

$$K^{a_1;a_n}(k) = a\binom{n}{k} \quad ; \quad K_{a_1;a_n}(k) = \frac{1}{a}\binom{n}{k}.$$

THEOREM 1. *If $M = \{..., a_{-2}, a_{-1}, a_0, a_1, a_2, ...\}$ is an ordered set of real positive numbers, we get*
(i)

$$
\begin{pmatrix} g_{a_i}^{a_j}(1,1) & g_{a_i}^{a_j}(1,2) \\ g_{a_i}^{a_j}(2,1) & g_{a_i}^{a_j}(2,2) \end{pmatrix} =
\begin{pmatrix}
\displaystyle\sum_{k=0}^{\left[\frac{i-j-1}{2}\right]} 3^k(-2)^{i-j-2k-1} K_{a_{i-1}}^{a_{j+1}}(2k) & -\displaystyle\sum_{k=0}^{\left[\frac{i-j-2}{2}\right]} 3^k(-2)^{i-j-2k-3} K^{a_{j+1};a_{i-1}}(2k+1) \\
-\displaystyle\sum_{k=0}^{\left[\frac{i-j-2}{2}\right]} 3^{k+1}(-2)^{i-j-2k-1} K_{a_{j+1};a_{i-1}}(2k+1) & \displaystyle\sum_{k=0}^{\left[\frac{i-j-1}{2}\right]} 3^k(-2)^{i-j-2k-1} K_{a_{j+1}}^{a_{i-1}}(2k)
\end{pmatrix}
$$

for $i - j > 1$,
(ii)

$$
\begin{pmatrix} g_{a_i}^{a_j}(1,1) & g_{a_i}^{a_j}(1,2) \\ g_{a_i}^{a_j}(2,1) & g_{a_i}^{a_j}(2,2) \end{pmatrix} = I \quad for \ i - j = 1,
$$

(iii)

$$
\begin{pmatrix} g_{a_i}^{a_j}(1,1) & g_{a_i}^{a_j}(1,2) \\ g_{a_i}^{a_j}(2,1) & g_{a_i}^{a_j}(2,2) \end{pmatrix} =
\begin{pmatrix}
\displaystyle\sum_{k=0}^{\left[\frac{j-i+1}{2}\right]} 3^k(-2)^{j-i-2k+1} K_{a_i}^{a_j}(2k) & \displaystyle\sum_{k=0}^{\left[\frac{j-i}{2}\right]} 3^k(-2)^{j-i-2k-1} K^{a_i;a_j}(2k+1) \\
\displaystyle\sum_{k=0}^{\left[\frac{j-i}{2}\right]} 3^{k+1}(-2)^{j-i-2k+1} K_{a_i;a_j}(2k+1) & \displaystyle\sum_{k=0}^{\left[\frac{j-i+1}{2}\right]} 3^k(-2)^{j-i-2k+1} K_{a_j}^{a_i}(2k)
\end{pmatrix}
$$

for $i - j < 1$.

Proof. By induction. ∎

LEMMA 2 will be useful in the following.

LEMMA 2. *For $x, y \in \mathbb{R}$ and $n \in \mathbb{N}$*

(i) $\displaystyle\sum_{i=0}^{\left[\frac{n}{2}\right]} \binom{n}{2i} x^{n-2i} \cdot y^{2i} = \frac{1}{2}\left((x+y)^n + (x-y)^n\right)$

(ii) $\displaystyle\sum_{i=0}^{\left[\frac{n-1}{2}\right]} \binom{n}{2i+1} x^{n-2i-1} \cdot y^{2i+1} = \frac{1}{2}\left((x+y)^n - (x-y)^n\right).$

Proof. (i)

$$(x+y)^n + (x-y)^n = \sum_{i=0}^{n} \binom{n}{i} x^{n-i} \left(y^i + (-y)^i \right)$$

$$(x+y)^n + (x-y)^n = 2 \sum_{i=0}^{\left[\frac{n}{2}\right]} \binom{n}{2i} x^{n-2i} \cdot y^{2i}.$$

(ii)

$$(x+y)^n - (x-y)^n = \sum_{i=0}^{n} \binom{n}{i} x^{n-i} \left(y^i - (-y)^i \right)$$

$$(x+y)^n - (x-y)^n = 2 \sum_{i=0}^{\left[\frac{n-1}{2}\right]} \binom{n}{2i+1} x^{n-2i-1} \cdot y^{2i+1}. \qquad \blacksquare$$

COROLLARY 2. *If all* a_k *are equal to a, we get*

(i)

$$\begin{pmatrix} g_{a_i}^{a_j}(1,1) & g_{a_i}^{a_j}(1,2) \\ g_{a_i}^{a_j}(2,1) & g_{a_i}^{a_j}(2,2) \end{pmatrix} = \begin{pmatrix} \frac{1}{2}\left((-2+\sqrt{3})^{i-j-1} + (-2-\sqrt{3})^{i-j-1} \right) & \frac{a}{4\sqrt{3}}\left((-2+\sqrt{3})^{i-j-1} - (-2-\sqrt{3})^{i-j-1} \right) \\ \frac{\sqrt{3}}{a}\left((-2+\sqrt{3})^{i-j-1} - (-2-\sqrt{3})^{i-j-1} \right) & \frac{1}{2}\left((-2+\sqrt{3})^{i-j-1} + (-2-\sqrt{3})^{i-j-1} \right) \end{pmatrix}$$

for $i - j > 1$,

(ii)

$$\begin{pmatrix} g_{a_i}^{a_j}(1,1) & g_{a_i}^{a_j}(1,2) \\ g_{a_i}^{a_j}(2,1) & g_{a_i}^{a_j}(2,2) \end{pmatrix} = I \text{ for } i-j=1,$$

(iii)

$$\begin{pmatrix} g_{a_i}^{a_j}(1,1) & g_{a_i}^{a_j}(1,2) \\ g_{a_i}^{a_j}(2,1) & g_{a_i}^{a_j}(2,2) \end{pmatrix} = \begin{pmatrix} \frac{1}{2}\left((-2+\sqrt{3})^{j-i+1} + (-2-\sqrt{3})^{j-i+1} \right) & -\frac{a}{4\sqrt{3}}\left((-2+\sqrt{3})^{j-i+1} - (-2-\sqrt{3})^{j-i+1} \right) \\ -\frac{\sqrt{3}}{a}\left((-2+\sqrt{3})^{j-i+1} - (-2-\sqrt{3})^{j-i+1} \right) & \frac{1}{2}\left((-2+\sqrt{3})^{j-i+1} + (-2-\sqrt{3})^{j-i+1} \right) \end{pmatrix}$$

for $i - j < 1$.

Proof. From THEOREM 1 immediately follows

(i)

$$\begin{pmatrix} g_{a_i}^{a_j}(1,1) & g_{a_i}^{a_j}(1,2) \\ g_{a_i}^{a_j}(2,1) & g_{a_i}^{a_j}(2,2) \end{pmatrix} = \begin{pmatrix} \sum_{k=0}^{\left[\frac{i-j-1}{2}\right]} 3^k (-2)^{i-j-2k-1} \binom{i-j-1}{2k} & -a \sum_{k=0}^{\left[\frac{i-j-2}{2}\right]} 3^k (-2)^{i-j-2k-3} \binom{i-j-1}{2k+1} \\ -\frac{1}{a} \sum_{k=0}^{\left[\frac{i-j-2}{2}\right]} 3^{k+1} (-2)^{i-j-2k-1} \binom{i-j-1}{2k+1} & \sum_{k=0}^{\left[\frac{i-j-1}{2}\right]} 3^k (-2)^{i-j-2k-1} \binom{i-j-1}{2k} \end{pmatrix}$$

for $i - j > 1$,

(ii)

$$\begin{pmatrix} g_{a_i}^{a_j}(1,1) & g_{a_i}^{a_j}(1,2) \\ g_{a_i}^{a_j}(2,1) & g_{a_i}^{a_j}(2,2) \end{pmatrix} = I \text{ for } i-j=1,$$

(iii)

$$\begin{pmatrix} g_{a_i}^{a_j}(1,1) & g_{a_i}^{a_j}(1,2) \\ g_{a_i}^{a_j}(2,1) & g_{a_i}^{a_j}(2,2) \end{pmatrix} = \begin{pmatrix} \sum_{k=0}^{\left[\frac{j-i+1}{2}\right]} 3^k(-2)^{j-i-2k+1}\binom{j-i+1}{2k} & a\sum_{k=0}^{\left[\frac{j-i}{2}\right]} 3^k(-2)^{j-i-2k-1}\binom{j-i+1}{2k+1} \\ \frac{1}{a}\sum_{k=0}^{\left[\frac{j-i}{2}\right]} 3^{k+1}(-2)^{j-i-2k+1}\binom{j-i+1}{2k+1} & \sum_{k=0}^{\left[\frac{j-i+1}{2}\right]} 3^k(-2)^{j-i-2k+1}\binom{j-i+1}{2k} \end{pmatrix}$$

for $i - j < 1$ and with LEMMA 2, the statement is ultimately obtained. ∎

With the help of DEFINITION 2, the four following lemmas can be proven:

LEMMA 3.

$$\begin{pmatrix} g_{a_i}^{a_j}(1,1) & g_{a_i}^{a_j}(1,2) \\ g_{a_i}^{a_j}(2,1) & g_{a_i}^{a_j}(2,2) \end{pmatrix} = \begin{pmatrix} -2g_{a_i}^{a_{j-1}}(1,1) - \frac{a_j}{2}g_{a_i}^{a_{j-1}}(2,1) & -2g_{a_i}^{a_{j-1}}(1,2) - \frac{a_j}{2}g_{a_i}^{a_{j-1}}(2,2) \\ -\frac{6}{a_j}g_{a_i}^{a_{j-1}}(1,1) - 2g_{a_i}^{a_{j-1}}(2,1) & -\frac{6}{a_j}g_{a_i}^{a_{j-1}}(1,2) - 2g_{a_i}^{a_{j-1}}(2,2) \end{pmatrix}.$$

LEMMA 4.

$$\begin{pmatrix} g_{a_i}^{a_j}(1,1) & g_{a_i}^{a_j}(1,2) \\ g_{a_i}^{a_j}(2,1) & g_{a_i}^{a_j}(2,2) \end{pmatrix} = \begin{pmatrix} -2g_{a_{i+1}}^{a_j}(1,1) - \frac{6}{a_i}g_{a_{i+1}}^{a_j}(1,2) & -\frac{a_i}{2}g_{a_{i+1}}^{a_j}(1,1) - 2g_{a_{i+1}}^{a_j}(1,2) \\ -2g_{a_{i+1}}^{a_j}(2,1) - \frac{6}{a_i}g_{a_{i+1}}^{a_j}(2,2) & -\frac{a_i}{2}g_{a_{i+1}}^{a_j}(2,1) - 2g_{a_{i+1}}^{a_j}(2,2) \end{pmatrix}.$$

LEMMA 5.

$$\begin{pmatrix} g_{a_i}^{a_{j-1}}(1,1) & g_{a_i}^{a_{j-1}}(1,2) \\ g_{a_i}^{a_{j-1}}(2,1) & g_{a_i}^{a_{j-1}}(2,2) \end{pmatrix} = \begin{pmatrix} -2g_{a_i}^{a_j}(1,1) + \frac{a_j}{2}g_{a_i}^{a_j}(2,1) & -2g_{a_i}^{a_j}(1,2) + \frac{a_j}{2}g_{a_i}^{a_j}(2,2) \\ \frac{6}{a_j}g_{a_i}^{a_j}(1,1) - 2g_{a_i}^{a_j}(2,1) & \frac{6}{a_j}g_{a_i}^{a_j}(1,2) - 2g_{a_i}^{a_j}(2,2) \end{pmatrix}.$$

LEMMA 6.

$$\begin{pmatrix} g_{a_{i+1}}^{a_j}(1,1) & g_{a_{i+1}}^{a_j}(1,2) \\ g_{a_{i+1}}^{a_j}(2,1) & g_{a_{i+1}}^{a_j}(2,2) \end{pmatrix} = \begin{pmatrix} -2g_{a_i}^{a_j}(1,1) + \frac{6}{a_i}g_{a_i}^{a_j}(1,2) & \frac{a_i}{2}g_{a_i}^{a_j}(1,1) - 2g_{a_i}^{a_j}(1,2) \\ -2g_{a_i}^{a_j}(2,1) + \frac{6}{a_i}g_{a_i}^{a_j}(2,2) & \frac{a_i}{2}g_{a_i}^{a_j}(2,1) - 2g_{a_i}^{a_j}(2,2) \end{pmatrix}.$$

LEMMA 7.

$$|G(a)| = 1 \qquad\qquad |G^{-1}(a)| = 1 .$$

LEMMA 8.

$$|G(a_j) \cdot G(a_{j-1}) \cdot \ldots \cdot G(a_i)| = 1 \quad j \geq i .$$

Proof. Using LEMMA 7:

$$|G(a_j) \cdot G(a_{j-1}) \cdot \ldots \cdot G(a_i)| = |G(a_j)| \cdot |G(a_{j-1})| \cdot \ldots \cdot |G(a_i)| = 1 . \quad ∎$$

LEMMA 9. *For $i \leq j \leq k$ it can be ascertained that*

$$\begin{vmatrix} g_{a_i}^{a_k}(1,1) & g_{a_{j+1}}^{a_k}(1,2) \\ g_{a_i}^{a_k}(2,1) & g_{a_{j+1}}^{a_k}(2,2) \end{vmatrix} = g_{a_i}^{a_j}(1,1)\,; \qquad \begin{vmatrix} g_{a_i}^{a_k}(2,1) & g_{a_{j+1}}^{a_k}(2,2) \\ g_{a_i}^{a_k}(1,1) & g_{a_{j+1}}^{a_k}(1,2) \end{vmatrix} = -g_{a_i}^{a_j}(1,1)\,;$$

$$\begin{vmatrix} g_{a_i}^{a_k}(1,2) & g_{a_{j+1}}^{a_k}(1,2) \\ g_{a_i}^{a_k}(2,2) & g_{a_{j+1}}^{a_k}(2,2) \end{vmatrix} = g_{a_i}^{a_j}(1,2)\,; \qquad \begin{vmatrix} g_{a_i}^{a_k}(2,2) & g_{a_{j+1}}^{a_k}(2,2) \\ g_{a_i}^{a_k}(1,2) & g_{a_{j+1}}^{a_k}(1,2) \end{vmatrix} = -g_{a_i}^{a_j}(1,2)\,;$$

$$\begin{vmatrix} g_{a_i}^{a_k}(2,1) & g_{a_{j+1}}^{a_k}(2,1) \\ g_{a_i}^{a_k}(1,1) & g_{a_{j+1}}^{a_k}(1,1) \end{vmatrix} = g_{a_i}^{a_j}(2,1)\,; \qquad \begin{vmatrix} g_{a_i}^{a_k}(1,1) & g_{a_{j+1}}^{a_k}(1,1) \\ g_{a_i}^{a_k}(2,1) & g_{a_{j+1}}^{a_k}(2,1) \end{vmatrix} = -g_{a_i}^{a_j}(2,1)\,;$$

$$\begin{vmatrix} g_{a_i}^{a_k}(2,2) & g_{a_{j+1}}^{a_k}(2,1) \\ g_{a_i}^{a_k}(1,2) & g_{a_{j+1}}^{a_k}(1,1) \end{vmatrix} = g_{a_i}^{a_j}(2,2)\,; \qquad \begin{vmatrix} g_{a_i}^{a_k}(1,2) & g_{a_{j+1}}^{a_k}(1,1) \\ g_{a_i}^{a_k}(2,2) & g_{a_{j+1}}^{a_k}(2,1) \end{vmatrix} = -g_{a_i}^{a_j}(2,2)\,.$$

Proof. The correctness of the statements of the second column is obtained from the correctness of the statements of the first column.

The statements of the first column are proven by induction.

For $j = k$, the theorem is true because $\begin{pmatrix} g_{a_{k+1}}^{a_k}(1,1) & g_{a_{k+1}}^{a_k}(1,2) \\ g_{a_{k+1}}^{a_k}(2,1) & g_{a_{k+1}}^{a_k}(2,2) \end{pmatrix} = I$ (DEFINITION 2 (ii)).

The proof of the first statement will be verified by way of example:

$$\begin{vmatrix} g_{a_i}^{a_k}(1,1) & g_{a_{j+1}}^{a_k}(1,2) \\ g_{a_i}^{a_k}(2,1) & g_{a_{j+1}}^{a_k}(2,2) \end{vmatrix} = g_{a_i}^{a_j}(1,1)\,.$$

Induction hypothesis

$$\begin{vmatrix} g_{a_i}^{a_k}(1,1) & g_{a_{j+2}}^{a_k}(1,2) \\ g_{a_i}^{a_k}(2,1) & g_{a_{j+2}}^{a_k}(2,2) \end{vmatrix} = g_{a_i}^{a_{j+1}}(1,1)\,; \qquad \begin{vmatrix} g_{a_i}^{a_k}(1,1) & g_{a_{j+2}}^{a_k}(1,1) \\ g_{a_i}^{a_k}(2,1) & g_{a_{j+2}}^{a_k}(2,1) \end{vmatrix} = -g_{a_i}^{a_{j+1}}(2,1)\,.$$

If LEMMA 4 is used, one obtains:

$$\begin{vmatrix} g_{a_i}^{a_k}(1,1) & g_{a_{j+1}}^{a_k}(1,2) \\ g_{a_i}^{a_k}(2,1) & g_{a_{j+1}}^{a_k}(2,2) \end{vmatrix} = \left| \begin{pmatrix} g_{a_i}^{a_k}(1,1) \\ g_{a_i}^{a_k}(2,1) \end{pmatrix} \quad -2\begin{pmatrix} g_{a_{j+2}}^{a_k}(1,2) \\ g_{a_{j+2}}^{a_k}(2,2) \end{pmatrix} - \frac{a_{j+1}}{2}\begin{pmatrix} g_{a_{j+2}}^{a_k}(1,1) \\ g_{a_{j+2}}^{a_k}(2,1) \end{pmatrix} \right|$$

$$\begin{vmatrix} g_{a_i}^{a_k}(1,1) & g_{a_{j+1}}^{a_k}(1,2) \\ g_{a_i}^{a_k}(2,1) & g_{a_{j+1}}^{a_k}(2,2) \end{vmatrix} = -2\begin{vmatrix} g_{a_i}^{a_k}(1,1) & g_{a_{j+2}}^{a_k}(1,2) \\ g_{a_i}^{a_k}(2,1) & g_{a_{j+2}}^{a_k}(2,2) \end{vmatrix} - \frac{a_{j+1}}{2}\begin{vmatrix} g_{a_i}^{a_k}(1,1) & g_{a_{j+2}}^{a_k}(1,1) \\ g_{a_i}^{a_k}(2,1) & g_{a_{j+2}}^{a_k}(2,1) \end{vmatrix}.$$

From the induction hypothesis and LEMMA 5 follow

$$\begin{vmatrix} g_{a_i}^{a_k}(1,1) & g_{a_{j+1}}^{a_k}(1,2) \\ g_{a_i}^{a_k}(2,1) & g_{a_{j+1}}^{a_k}(2,2) \end{vmatrix} = -2g_{a_i}^{a_{j+1}}(1,1) + \frac{a_{j+1}}{2}g_{a_i}^{a_{j+1}}(2,1) = g_{a_i}^{a_j}(1,1)\,. \qquad \blacksquare$$

DEFINITION 4. *It is assumed that* $\ldots, \Delta_{-2}, \Delta_{-1}, \Delta_0, \Delta_1, \Delta_2, \ldots$ *are positive real numbers with the quality of periodicity*

$$\Delta_i = \Delta_{i+n}\,; \quad n > 1\,; \quad n \in I\!N\,.$$

DEFINITION 5.

$$v_n = 2 - g_{\Delta_l}^{\Delta_{l+n-1}}(1,1) - g_{\Delta_l}^{\Delta_{l+n-1}}(2,2)\,.$$

THEOREM 2. v_n *is independent of l, since*

$$v_n = 2 - g_{\Delta_0}^{\Delta_{n-1}}(1,1) - g_{\Delta_0}^{\Delta_{n-1}}(2,2)$$

is true.

Proof. Because of DEFINITION 2 we can write

$$G(\Delta_{l+n-1})G(\Delta_{l+n-2})...G(\Delta_l) = \begin{pmatrix} g_{\Delta_n}^{\Delta_{l+n-1}}(1,1) & g_{\Delta_n}^{\Delta_{l+n-1}}(1,2) \\ g_{\Delta_n}^{\Delta_{l+n-1}}(2,1) & g_{\Delta_n}^{\Delta_{l+n-1}}(2,2) \end{pmatrix} \begin{pmatrix} g_{\Delta_l}^{\Delta_{n-1}}(1,1) & g_{\Delta_l}^{\Delta_{n-1}}(1,2) \\ g_{\Delta_l}^{\Delta_{n-1}}(2,1) & g_{\Delta_l}^{\Delta_{n-1}}(2,2) \end{pmatrix}$$

and, with this, the following is obtained due to the periodicity of Δ_l:

$$v_n = 2 - g_{\Delta_0}^{\Delta_{l-1}}(1,1)g_{\Delta_l}^{\Delta_{n-1}}(1,1) - g_{\Delta_0}^{\Delta_{l-1}}(1,2)g_{\Delta_l}^{\Delta_{n-1}}(2,1) - g_{\Delta_0}^{\Delta_{l-1}}(2,1)g_{\Delta_l}^{\Delta_{n-1}}(1,2) - g_{\Delta_0}^{\Delta_{l-1}}(2,2)g_{\Delta_l}^{\Delta_{n-1}}(2,2).$$

Because of LEMMA 9, $g_{\Delta_l}^{\Delta_{l-1}}$ can be represented by $g_{\Delta_0}^{\Delta_{n-1}}$ and $g_{\Delta_l}^{\Delta_{n-1}}$. It follows from this that

$$v_n = 2 - g_{\Delta_0}^{\Delta_{n-1}}(1,1) \cdot \begin{vmatrix} g_{\Delta_l}^{\Delta_{n-1}}(1,1) & g_{\Delta_l}^{\Delta_{n-1}}(1,2) \\ g_{\Delta_l}^{\Delta_{n-1}}(2,1) & g_{\Delta_l}^{\Delta_{n-1}}(2,2) \end{vmatrix} - g_{\Delta_0}^{\Delta_{n-1}}(2,2) \cdot \begin{vmatrix} g_{\Delta_l}^{\Delta_{n-1}}(1,1) & g_{\Delta_l}^{\Delta_{n-1}}(1,2) \\ g_{\Delta_l}^{\Delta_{n-1}}(2,1) & g_{\Delta_l}^{\Delta_{n-1}}(2,2) \end{vmatrix} .$$

LEMMA 8 gives the statement. ∎

Through THEOREM 2 and THEOREM 1, one obtains:

COROLLARY 3.

$$v_n = 2 - \sum_{i=0}^{\left[\frac{n}{2}\right]} 3^i (-2)^{n-2i} K_{\Delta_0}^{\Delta_{n-1}}(2i) - \sum_{i=0}^{\left[\frac{n}{2}\right]} 3^i (-2)^{n-2i} K_{\Delta_{n-1}}^{\Delta_0}(2i).$$

COROLLARY 4. *With* $\Delta_i = \Delta$ $(i = 0, 1, ..., n-1)$ *the* v_n *(*$n = 1, 2, 3, ...$*) are independent of* Δ_i, *only depend on n and are equal to*

$$v_n = 2 + (-1)^{n-1}\left((2+\sqrt{3})^n + (2-\sqrt{3})^n\right)$$

Proof. Because of THEOREM 2 and COROLLARY 2 we can write

$$v_n = 2 - 2 \cdot \frac{1}{2}\left((-2+\sqrt{3})^n + (-2-\sqrt{3})^n\right).$$

This yields the statement. ∎

DEFINITION 6.

$$h'_{r,l,n} = \frac{\Delta_l}{\Delta_{l+r}}\left(g_{\Delta_l}^{\Delta_{l+r-1}}(2,2) + g_{\Delta_{l+r+1}}^{\Delta_{l+n-1}}(1,1)\right) + \frac{2\Delta_l}{\Delta_{l+r}^2}\left(g_{\Delta_l}^{\Delta_{l+r-1}}(1,2) + g_{\Delta_{l+r+1}}^{\Delta_{l+n-1}}(1,2)\right)$$

and

$$h''_{r,l,n} = -\frac{\Delta_l}{\Delta_{l+r}}\left(g_{\Delta_{l+1}}^{\Delta_{l+r-1}}(2,2) + g_{\Delta_{l+r+1}}^{\Delta_{l+n}}(1,1)\right) - \frac{2\Delta_l}{\Delta_{l+r}^2}\left(g_{\Delta_{l+1}}^{\Delta_{l+r-1}}(1,2) + g_{\Delta_{l+r+1}}^{\Delta_{l+n}}(1,2)\right).$$

THEOREM 3. $h'_{r,l,n}$ *and* $h''_{r,l,n}$ *can be expressed in terms of each other as*

$$h'_{r,l,n} = -\frac{\Delta_l}{\Delta_{l-1}}h''_{r+1,l-1,n}$$

and

$$h''_{r,l,n} = -\frac{\Delta_l}{\Delta_{l+1}}h'_{r-1,l+1,n}$$

is true.

Proof. Because of DEFINITION 6

$$h'_{r-1,l+1,n} = -\frac{\Delta_{l+1}}{\Delta_l}\left(-\frac{\Delta_l}{\Delta_{l+r}}\left(g^{\Delta_{l+r-1}}_{\Delta_{l+1}}(2,2)+g^{\Delta_{l+n}}_{\Delta_{l+r+1}}(1,1)\right)-\frac{2\Delta_l}{\Delta_{l+r}^2}\left(g^{\Delta_{l+r-1}}_{\Delta_{l+1}}(1,2)+g^{\Delta_{l+n}}_{\Delta_{l+r+1}}(1,2)\right)\right)$$

$$h'_{r-1,l+1,n} = -\frac{\Delta_{l+1}}{\Delta_l}h''_{r,l,n}$$

and with this both statements are valid. ∎

With THEOREM 1 we get

COROLLARY 5.

$$h'_{r,l,n} = \frac{\Delta_l}{\Delta_{l+r}}\left(\sum_{i=0}^{\left[\frac{r}{2}\right]}3^i(-2)^{r-2i}K^{\Delta_l}_{\Delta_{l+r-1}}(2i)+\sum_{i=0}^{\left[\frac{n-r-1}{2}\right]}3^i(-2)^{n-r-2i-1}K^{\Delta_{l+n-1}}_{\Delta_{l+r+1}}(2i)\right)-$$

$$\frac{\Delta_l}{\Delta_{l+r}^2}\left(\sum_{i=0}^{\left[\frac{r-1}{2}\right]}3^i(-2)^{r-2i-1}K^{\Delta_l;\Delta_{l+r-1}}(2i+1)+\sum_{i=0}^{\left[\frac{n-r-2}{2}\right]}3^i(-2)^{n-r-2i-2}K^{\Delta_{l+r+1};\Delta_{l+n-1}}(2i+1)\right)$$

and

$$h''_{r,l,n} = \delta_{0,r}-\frac{\Delta_l}{\Delta_{l+r}}\left(\sum_{i=0}^{\left[\frac{r-1}{2}\right]}3^i(-2)^{r-2i-1}K^{\Delta_{l+1}}_{\Delta_{l+r-1}}(2i)+\sum_{i=0}^{\left[\frac{n-r}{2}\right]}3^i(-2)^{n-r-2i}K^{\Delta_{l+n}}_{\Delta_{l+r+1}}(2i)\right)+$$

$$\frac{\Delta_l}{\Delta_{l+r}^2}\left(\sum_{i=0}^{\left[\frac{r-2}{2}\right]}3^i(-2)^{r-2i-2}K^{\Delta_{l+1};\Delta_{l+r-1}}(2i+1)+\sum_{i=0}^{\left[\frac{n-r-1}{2}\right]}3^i(-2)^{n-r-2i-1}K^{\Delta_{l+r+1};\Delta_{l+n}}(2i+1)\right).$$

COROLLARY 6. *With $\Delta_i = \Delta$ $(i = 0, 1, ..., n-1)$ (equidistant parameters)*

$$h'_{r,l,n} = \frac{1}{6}\sqrt{3}\left(\left(-\frac{1}{2}\right)^r\left((\sqrt{3}+1)^{2r+1}+(\sqrt{3}-1)^{2r+1}\right)+\left(-\frac{1}{2}\right)^{n-r-1}\left((\sqrt{3}+1)^{2n-2r-1}+(\sqrt{3}-1)^{2n-2r-1}\right)\right)$$

and

$$h''_{r,l,n} = -\frac{1}{6}\sqrt{3}\left(\left(-\frac{1}{2}\right)^{r-1}\left((\sqrt{3}+1)^{2r-1}+(\sqrt{3}-1)^{2r-1}\right)+\left(-\frac{1}{2}\right)^{n-r}\left((\sqrt{3}+1)^{2n-2r+1}+(\sqrt{3}-1)^{2n-2r+1}\right)\right)$$

are true. Thus, $h'_{r,l,n}$ and $h''_{r,l,n}$ neither depend on l nor on Δ_k.

Proof. Because of DEFINITION 6

$$h'_{r,l,n} = \left(g^{\Delta_{l+r-1}}_{\Delta_l}(2,2)+g^{\Delta_{l+n-1}}_{\Delta_{l+r+1}}(1,1)\right)+\frac{2}{\Delta}\left(g^{\Delta_{l+r-1}}_{\Delta_l}(1,2)+g^{\Delta_{l+n-1}}_{\Delta_{l+r+1}}(1,2)\right).$$

Using COROLLARY 2 (iii), we can write

$$h'_{r,l,n} = \left(\frac{1}{2}-\frac{1}{2\sqrt{3}}\right)(-2+\sqrt{3})^r+\left(\frac{1}{2}+\frac{1}{2\sqrt{3}}\right)(-2-\sqrt{3})^r+$$

$$\left(\frac{1}{2}-\frac{1}{2\sqrt{3}}\right)(-2+\sqrt{3})^{n-r-1}+\left(\frac{1}{2}+\frac{1}{2\sqrt{3}}\right)(-2-\sqrt{3})^{n-r-1},$$

$$h'_{r,l,n} = \left(-\frac{1}{2}\right)^r \left(\frac{\sqrt{3}-1}{2\sqrt{3}}\right)(4-2\sqrt{3})^r + \left(-\frac{1}{2}\right)^r \left(\frac{\sqrt{3}+1}{2\sqrt{3}}\right)(4+2\sqrt{3})^r +$$

$$\left(-\frac{1}{2}\right)^{n-r-1} \left(\frac{\sqrt{3}-1}{2\sqrt{3}}\right)(4-2\sqrt{3})^{n-r-1} + \left(-\frac{1}{2}\right)^{n-r-1} \left(\frac{\sqrt{3}+1}{2\sqrt{3}}\right)(4+2\sqrt{3})^{n-r-1}$$

and finally

$$h'_{r,l,n} = \left(-\frac{1}{2}\right)^r \frac{\sqrt{3}}{6}(\sqrt{3}-1)^{2r+1} + \left(-\frac{1}{2}\right)^r \frac{\sqrt{3}}{6}(\sqrt{3}+1)^{2r+1} +$$

$$\left(-\frac{1}{2}\right)^{n-r-1} \frac{\sqrt{3}}{6}(\sqrt{3}-1)^{2n-2r-1} + \left(-\frac{1}{2}\right)^{n-r-1} \frac{\sqrt{3}}{6}(\sqrt{3}+1)^{2n-2r-1}.$$

The statement for $h''_{r,l,n}$ immediately follows from $h''_{r,l,n} = -h'_{r-1,l+1,n}$ (THEOREM 3).   ∎

THEOREM 4. *The $h'_{r,l,n}$ and the $h''_{r,l,n}$ are periodical with respect to l, for*

$$h'_{r,l+n,n} = h'_{r,l,n}$$

and

$$h''_{r,l+n,n} = h''_{r,l,n}$$

are true.

Proof. This statement follows directly from the periodicity of the Δ_l in DEFINITION 6.   ∎

THEOREM 5. *It describes the possibility of calculating $h'_{r,l,n}$ and $h''_{r,l,n}$ iteratively, for*

$$\frac{\Delta_l}{\Delta_{l-1}} h'_{r+1,l-1,n} + 2\left(1+\frac{\Delta_{l-1}}{\Delta_l}\right) h'_{r,l,n} + \frac{\Delta_{l-1}}{\Delta_{l+1}} h'_{r-1,l+1,n} = 0$$

and

$$\frac{\Delta_{l+1}}{\Delta_{l-1}} h''_{r+1,l-1,n} + 2\left(1+\frac{\Delta_{l+1}}{\Delta_l}\right) h''_{r,l,n} + \frac{\Delta_l}{\Delta_{l+1}} h''_{r-1,l+1,n} = 0$$

are true.

Proof. With DEFINITION 6 we get

$$\frac{\Delta_l}{\Delta_{l-1}} h'_{r+1,l-1,n} + 2\left(1+\frac{\Delta_{l-1}}{\Delta_l}\right) h'_{r,l,n} + \frac{\Delta_{l-1}}{\Delta_{l+1}} h'_{r-1,l+1,n} =$$

$$\frac{\Delta_l}{\Delta_{l-1}} \left(\frac{\Delta_{l-1}}{\Delta_{l+r}} \left(g_{\Delta_{l-1}}^{\Delta_{l+r-1}}(2,2) + g_{\Delta_{l+r+1}}^{\Delta_{l+r-2}}(1,1) \right) + \frac{2\Delta_{l-1}}{\Delta_{l+r}^2} \left(g_{\Delta_{l-1}}^{\Delta_{l+r-1}}(1,2) + g_{\Delta_{l+r+1}}^{\Delta_{l+r-2}}(1,2) \right) \right) +$$

$$2\left(1+\frac{\Delta_{l-1}}{\Delta_l}\right) \left(\frac{\Delta_l}{\Delta_{l+r}} \left(g_{\Delta_l}^{\Delta_{l+r-1}}(2,2) + g_{\Delta_{l+r+1}}^{\Delta_{l+n-1}}(1,1) \right) + \frac{2\Delta_l}{\Delta_{l+r}^2} \left(g_{\Delta_l}^{\Delta_{l+r-1}}(1,2) + g_{\Delta_{l+r+1}}^{\Delta_{l+n-1}}(1,2) \right) \right) +$$

$$\frac{\Delta_{l-1}}{\Delta_{l+1}} \left(\frac{\Delta_{l+1}}{\Delta_{l+r}} \left(g_{\Delta_{l+1}}^{\Delta_{l+r-1}}(2,2) + g_{\Delta_{l+r+1}}^{\Delta_{l+n}}(1,1) \right) + \frac{2\Delta_{l+1}}{\Delta_{l+r}^2} \left(g_{\Delta_{l+1}}^{\Delta_{l+r-1}}(1,2) + g_{\Delta_{l+r+1}}^{\Delta_{l+n}}(1,2) \right) \right).$$

Using LEMMA 4, LEMMA 5, LEMMA 6 and LEMMA 3, the right side of the above equation can be expressed in terms of $g_{\Delta_l}^{\Delta_{l+r-1}}$ und $g_{\Delta_{l+r+1}}^{\Delta_{l+n-1}}$. This yields the form

$$\frac{\Delta_l}{\Delta_{l-1}}h'_{r+1,l-1,n} + 2\left(1+\frac{\Delta_{l-1}}{\Delta_l}\right)h'_{r,l,n} + \frac{\Delta_{l-1}}{\Delta_{l+1}}h'_{r-1,l+1,n} = \alpha_1 g_{\Delta_l}^{\Delta_{l+r-1}}(1,1) + \beta_1 g_{\Delta_{l+r+1}}^{\Delta_{l+n-1}}(1,1) + \alpha_2 g_{\Delta_l}^{\Delta_{l+r-1}}(1,2) +$$

$$\beta_2 g_{\Delta_{l+r-1}}^{\Delta_{l+n-1}}(1,2) + \alpha_3 g_{\Delta_l}^{\Delta_{l+r-1}}(2,1) + \beta_3 g_{\Delta_{l+r+1}}^{\Delta_{l+n-1}}(2,1) + \alpha_4 g_{\Delta_l}^{\Delta_{l+r-1}}(2,2) + \beta_4 g_{\Delta_{l+r+1}}^{\Delta_{l+n-1}}(2,2),$$

in which $\alpha_1 = \beta_1 = \alpha_2 = \beta_2 = \alpha_3 = \beta_3 = \alpha_4 = \beta_4 = 0$.

If one uses THEOREM 3 for the first statement of the theorem, then

$$-\frac{\Delta_l}{\Delta_{l-2}}h''_{r+2,l-2,n} - 2\left(\frac{\Delta_l}{\Delta_{l-1}}+1\right)h''_{r+1,l-1,n} - \frac{\Delta_{l-1}}{\Delta_l}h''_{r,l,n} = 0.$$

Multiplying this equation with -1 and with suitable index transformations for l and r: ∎

THEOREM 6.

$$\frac{h'_{n-1,l,n}}{v_n} = \frac{h'_{-1,l,n}}{v_n} + \frac{\Delta_l}{\Delta_{l-1}}$$

and

$$\frac{h''_{n-1,l,n}}{v_n} = \frac{h''_{-1,l,n}}{v_n} + \frac{\Delta_l^2}{\Delta_{l-1}^2} + \frac{2\Delta_l}{\Delta_{l-1}}$$

are true.

Proof. With DEFINITION 6 and the periodicity of Δ_l:

$$h'_{n-1,l,n} = \frac{\Delta_l}{\Delta_{l-1}}\left(g_{\Delta_l}^{\Delta_{l+n-2}}(2,2) + g_{\Delta_{l+n}}^{\Delta_{l+n-1}}(1,1)\right) + \frac{2\Delta_l}{\Delta_{l-1}^2}\left(g_{\Delta_l}^{\Delta_{l+n-2}}(1,2) + g_{\Delta_{l+n}}^{\Delta_{l+n-1}}(1,2)\right).$$

We use DEFINITION 2 (ii) for $g_{\Delta_{l+n}}^{\Delta_{l+n-1}}$ and transform $g_{\Delta_l}^{\Delta_{l+n-2}}$ using LEMMA 5 into expressions containing $g_{\Delta_l}^{\Delta_{l+n-1}}$. From this follows:

$$h'_{n-1,l,n} = \frac{2\Delta_l}{\Delta_{l-1}^2}g_{\Delta_l}^{\Delta_{l+n-1}}(1,2) - \frac{\Delta_l}{\Delta_{l-1}}g_{\Delta_l}^{\Delta_{l+n-1}}(2,2) + \frac{\Delta_l}{\Delta_{l-1}}.$$

Consequently:

$$h'_{n-1,l,n} = \frac{\Delta_l}{\Delta_{l-1}}\left(-2 + g_{\Delta_l}^{\Delta_{l+n-1}}(1,1)\right) + \frac{2\Delta_l}{\Delta_{l-1}^2}\left(\frac{\Delta_{l-1}}{2} + g_{\Delta_l}^{\Delta_{l+n-1}}(1,2)\right) + \frac{\Delta_l}{\Delta_{l-1}}\left(2 - g_{\Delta_l}^{\Delta_{l+n-1}}(1,1) - g_{\Delta_l}^{\Delta_{l+n-1}}(2,2)\right)$$

and finally with DEFINITION 6 and DEFINITION 5:

$$h'_{n-1,l,n} = h'_{-1,l,n} + \frac{\Delta_l}{\Delta_{l-1}}v_n.$$

With DEFINITION 6 and the periodicity of Δ_l we get

$$h''_{n-1,l,n} = -\frac{\Delta_l}{\Delta_{l-1}}\left(g_{\Delta_{l+1}}^{\Delta_{l+n-2}}(2,2) + g_{\Delta_{l+n}}^{\Delta_{l+n}}(1,1)\right) - \frac{2\Delta_l}{\Delta_{l-1}^2}\left(g_{\Delta_{l+1}}^{\Delta_{l+n-2}}(1,2) + g_{\Delta_{l+n}}^{\Delta_{l+n}}(1,2)\right).$$

We use DEFINITION 2 (iii) for $g_{\Delta_{l+n}}^{\Delta_{l+n}}$ and transform $g_{\Delta_{l+1}}^{\Delta_{l+n-2}}$ using LEMMA 5 and LEMMA 6 into expressions containing $g_{\Delta_l}^{\Delta_{l+n-1}}$. This yields:

$$h''_{n-1,l,n} = -\frac{\Delta_l^2}{\Delta_{l-1}^2}g_{\Delta_l}^{\Delta_{l+n-1}}(1,1) + \frac{\Delta_l^2}{2\Delta_{l-1}}g_{\Delta_l}^{\Delta_{l+n-1}}(2,1) + \frac{4\Delta_l}{\Delta_{l-1}^2}g_{\Delta_l}^{\Delta_{l+n-1}}(1,2) - \frac{2\Delta_l}{\Delta_{l-1}}g_{\Delta_l}^{\Delta_{l+n-1}}(2,2) + \frac{2\Delta_l}{\Delta_{l-1}} + \frac{\Delta_l^2}{\Delta_{l-1}^2}.$$

It follows that:

$$h''_{n-1,l,n} = -\frac{\Delta_l}{\Delta_{l-1}}\left(\frac{3\Delta_l}{\Delta_{l-1}} + 4 - 2g_{\Delta_l}^{\Delta_{l+n-1}}(1,1) - \frac{\Delta_l}{2}g_{\Delta_l}^{\Delta_{l+n-1}}(2,1)\right) -$$

$$\frac{2\Delta_l}{\Delta_{l-1}^2}\left(-\Delta_l - \Delta_{l-1} - 2g_{\Delta_l}^{\Delta_{l+n-1}}(1,2) - \frac{\Delta_l}{2}g_{\Delta_l}^{\Delta_{l+n-1}}(2,2)\right) +$$

$$\frac{\Delta_l^2}{\Delta_{l-1}^2}\left(2 - g_{\Delta_l}^{\Delta_{l+n-1}}(1,1) - g_{\Delta_l}^{\Delta_{l+n-1}}(2,2)\right) + \frac{2\Delta_l}{\Delta_{l-1}}\left(2 - g_{\Delta_l}^{\Delta_{l+n-1}}(1,1) - g_{\Delta_l}^{\Delta_{l+n-1}}(2,2)\right)$$

and finally with DEFINITION 6, LEMMA 3 and DEFINITION 5

$$h''_{n-1,l,n} = h''_{-1,l,n} + \left(\frac{\Delta_l^2}{\Delta_{l-1}^2} + \frac{2\Delta_l}{\Delta_{l-1}}\right)v_n. \quad \blacksquare$$

Because of the C^1-continuity (cf. Farin [1, (9.1)])

$$\mathbf{x}_l = \frac{\Delta_l}{\Delta_{l-1} + \Delta_l}\mathbf{b}_{3l-1} + \frac{\Delta_{l-1}}{\Delta_{l-1} + \Delta_l}\mathbf{b}_{3l+1}$$

(1)
$$\mathbf{b}_{3l+1} = -\frac{\Delta_l}{\Delta_{l-1}}\mathbf{b}_{3l-1} + \frac{\Delta_{l-1} + \Delta_l}{\Delta_{l-1}}\mathbf{x}_l .$$

Also because of the C^2-continuity (cf. Farin [1, (7.8), (7.9)])

(2)
$$\mathbf{b}_{3l-1} = \frac{\Delta_l}{\Delta_{l-1} + \Delta_l}\mathbf{b}_{3l-2} + \frac{\Delta_{l-1}}{\Delta_{l-1} + \Delta_l}\mathbf{d}_{l-1} ;$$

$$\mathbf{b}_{3l+1} = \frac{\Delta_l}{\Delta_{l-1} + \Delta_l}\mathbf{d}_{l-1} + \frac{\Delta_{l-1}}{\Delta_{l-1} + \Delta_l}\mathbf{b}_{3l+2} .$$

If the next to last equation is multiplied by $\Delta_l(\Delta_{l-1} + \Delta_l)$ and the last equation by $\Delta_{l-1}(\Delta_{l-1} + \Delta_l)$, solving the two equations for $\Delta_{l-1}\Delta_l\mathbf{d}_{l-1}$ and equating the right sides, one gets:

$$\Delta_l(\Delta_{l-1} + \Delta_l)\mathbf{b}_{3l-1} - \Delta_l^2\mathbf{b}_{3l-2} = \Delta_{l-1}(\Delta_{l-1} + \Delta_l)\mathbf{b}_{3l+1} - \Delta_{l-1}^2\mathbf{b}_{3l+2} .$$

If the right side of (1) (condition for C^1-continuity) is plugged into the last equation, one finally gets:

(3)
$$\mathbf{b}_{3l+2} = \frac{\Delta_l^2}{\Delta_{l-1}^2}\mathbf{b}_{3l-2} - \frac{2\Delta_l(\Delta_{l-1} + \Delta_l)}{\Delta_{l-1}^2}\mathbf{b}_{3l-1} + \frac{(\Delta_{l-1} + \Delta_l)^2}{\Delta_{l-1}^2}\mathbf{x}_l .$$

THEOREM 7. *For a* $(\mathbf{x}_n - \mathbf{x}_0)$-*periodic cubic B-spline curve, which interpolates in given data points* $\mathbf{x}_0, \mathbf{x}_1, \ldots, \mathbf{x}_{n-1}, \mathbf{x}_n = \mathbf{x}_0 + \mathbf{p}$, *one gets the Bézier points*

$$\mathbf{b}_{3l+1} = \mathbf{x}_l + \frac{1}{v_n}\sum_{r=0}^{n-1} h'_{r,l,n}\Delta\mathbf{x}_{l+r}$$

and

$$\mathbf{b}_{3l+2} = \mathbf{x}_l + \frac{1}{v_n}\sum_{r=0}^{n-1} h''_{r,l,n}\Delta\mathbf{x}_{l+r} .$$

Proof. The following properties a), b) and c) will be proven successively.

a)

(1)
$$\mathbf{b}_{3l+1} = -\frac{\Delta_l}{\Delta_{l-1}}\mathbf{b}_{3l-1} + \frac{\Delta_{l-1}+\Delta_l}{\Delta_{l-1}}\mathbf{x}_l$$

b)

(3)
$$\mathbf{b}_{3l+2} = \frac{\Delta_l^2}{\Delta_{l-1}^2}\mathbf{b}_{3l-2} - \frac{2\Delta_l(\Delta_{l-1}+\Delta_l)}{\Delta_{l-1}^2}\mathbf{b}_{3l-1} + \frac{(\Delta_{l-1}+\Delta_l)^2}{\Delta_{l-1}^2}\mathbf{x}_l$$

and
c)

$$\mathbf{b}_{3l+3n+1} = \mathbf{b}_{3l+1} + \mathbf{p}$$

and

$$\mathbf{b}_{3l+3n+2} = \mathbf{b}_{3l+2} + \mathbf{p}.$$

a) With the statement of THEOREM 6 and a suitable index transformation for r:

$$-\frac{\Delta_l}{\Delta_{l-1}}\mathbf{b}_{3l-1} + \frac{\Delta_{l-1}+\Delta_l}{\Delta_{l-1}}\mathbf{x}_l = \mathbf{x}_l + \frac{\Delta_l}{\Delta_{l-1}}\Delta\mathbf{x}_{l-1} - \frac{\Delta_l}{\Delta_{l-1}v_n}\sum_{r=-1}^{n-2}h''_{r+1,l-1,n}\Delta\mathbf{x}_{l+r}.$$

With THEOREM 3:

$$-\frac{\Delta_l}{\Delta_{l-1}}\mathbf{b}_{3l-1} + \frac{\Delta_{l-1}+\Delta_l}{\Delta_{l-1}}\mathbf{x}_l = \mathbf{x}_l + \frac{\Delta_l}{\Delta_{l-1}}\Delta\mathbf{x}_{l-1} + \frac{1}{v_n}h'_{-1,l,n}\Delta\mathbf{x}_{l-1} + \frac{1}{v_n}\sum_{r=0}^{n-2}h'_{r,l,n}\Delta\mathbf{x}_{l+r}$$

and with THEOREM 6:

$$-\frac{\Delta_l}{\Delta_{l-1}}\mathbf{b}_{3l-1} + \frac{\Delta_{l-1}+\Delta_l}{\Delta_{l-1}}\mathbf{x}_l = \mathbf{x}_l + \frac{1}{v_n}\sum_{r=0}^{n-2}h'_{r,l,n}\Delta\mathbf{x}_{l+r} + \frac{1}{v_n}h'_{n-1,l,n}\Delta\mathbf{x}_{l-1}$$

and finally the statement.

b) With the statement of THEOREM 6 and a suitable index transformation for r we get

$$\frac{\Delta_l^2}{\Delta_{l-1}^2}\mathbf{b}_{3l-2} - \frac{2\Delta_l(\Delta_{l-1}+\Delta_l)}{\Delta_{l-1}^2}\mathbf{b}_{3l-1} + \frac{(\Delta_{l-1}+\Delta_l)^2}{\Delta_{l-1}^2}\mathbf{x}_l = \frac{\Delta_l^2}{\Delta_{l-1}^2}\mathbf{x}_{l-1} - \frac{2\Delta_l}{\Delta_{l-1}}\mathbf{x}_{l-1} - \frac{2\Delta_l^2}{\Delta_{l-1}^2}\mathbf{x}_{l-1} +$$

$$\frac{\Delta_{l-1}^2}{\Delta_{l-1}^2}\mathbf{x}_l + \frac{2\Delta_l}{\Delta_{l-1}}\mathbf{x}_l + \frac{\Delta_l^2}{\Delta_{l-1}^2}\mathbf{x}_l + \frac{\Delta_l^2}{\Delta_{l-1}^2 v_n}\sum_{r=-1}^{n-2}h''_{r+1,l-1,n}\Delta\mathbf{x}_{l+r} - \frac{2\Delta_l(\Delta_{l-1}+\Delta_l)}{\Delta_{l-1}^2 v_n}\sum_{r=-1}^{n-2}h''_{r+1,l-1,n}\Delta\mathbf{x}_{l+r}.$$

THEOREM 3 applied to $h'_{r+1,l-1,n}$, yields:

$$\frac{\Delta_l^2}{\Delta_{l-1}^2}\mathbf{b}_{3l-2} - \frac{2\Delta_l(\Delta_{l-1}+\Delta_l)}{\Delta_{l-1}^2}\mathbf{b}_{3l-1} + \frac{(\Delta_{l-1}+\Delta_l)^2}{\Delta_{l-1}^2}\mathbf{x}_l = \mathbf{x}_l + \frac{2\Delta_l}{\Delta_{l-1}}\Delta\mathbf{x}_{l-1} + \frac{\Delta_l^2}{\Delta_{l-1}^2}\Delta\mathbf{x}_{l-1} -$$

$$\frac{\Delta_l}{\Delta_{l-1}v_n}\sum_{r=-1}^{n-2}\left(\frac{\Delta_l}{\Delta_{l-2}}h''_{r+2,l-2,n} + 2\left(1+\frac{\Delta_l}{\Delta_{l-1}}\right)h''_{r+1,l-1,n}\right)\Delta\mathbf{x}_{l+r}.$$

Because of THEOREM 5:

$$\frac{\Delta_l^2}{\Delta_{l-1}^2}\mathbf{b}_{3l-2} - \frac{2\Delta_l(\Delta_{l-1}+\Delta_l)}{\Delta_{l-1}^2}\mathbf{b}_{3l-1} + \frac{(\Delta_{l-1}+\Delta_l)^2}{\Delta_{l-1}^2}\mathbf{x}_l = \mathbf{x}_l + \frac{2\Delta_l}{\Delta_{l-1}}\Delta\mathbf{x}_{l-1} + \frac{\Delta_l^2}{\Delta_{l-1}^2}\Delta\mathbf{x}_{l-1} -$$

$$\frac{\Delta_l}{\Delta_{l-1}v_n}\left(-\frac{\Delta_{l-1}}{\Delta_l}h''_{-1,l,n}\right)\Delta\mathbf{x}_{l-1} - \frac{\Delta_l}{\Delta_{l-1}v_n}\sum_{r=0}^{n-2}\left(-\frac{\Delta_{l-1}}{\Delta_l}h''_{r,l,n}\right)\Delta\mathbf{x}_{l+r}.$$

From THEOREM 6 follows:

$$\frac{\Delta_l^2}{\Delta_{l-1}^2}\mathbf{b}_{3l-2} - \frac{2\Delta_l(\Delta_{l-1}+\Delta_l)}{\Delta_{l-1}^2}\mathbf{b}_{3l-1} + \frac{(\Delta_{l-1}+\Delta_l)^2}{\Delta_{l-1}^2}\mathbf{x}_l = \mathbf{x}_l + \frac{h''_{n-1,l,n}}{v_n}\Delta\mathbf{x}_{l-1} + \frac{1}{v_n}\sum_{r=0}^{n-2}h''_{r,l,n}\Delta\mathbf{x}_{l+r}$$

and finally the statement.

c)

$$\mathbf{b}_{3(l+n)+1} = \mathbf{x}_{l+n} + \frac{1}{v_n}\sum_{r=0}^{n-1} h'_{r,l,n}\Delta\mathbf{x}_{l+r} = \mathbf{x}_l + \mathbf{p} + \frac{1}{v_n}\sum_{r=0}^{n-1} h'_{r,l,n}\Delta\mathbf{x}_{l+r} = \mathbf{b}_{3l+1} + \mathbf{p}$$

or

$$\mathbf{b}_{3(l+n)+2} = \mathbf{x}_{l+n} + \frac{1}{v_n}\sum_{r=0}^{n-1} h''_{r,l,n}\Delta\mathbf{x}_{l+r} = \mathbf{x}_l + \mathbf{p} + \frac{1}{v_n}\sum_{r=0}^{n-1} h''_{r,l,n}\Delta\mathbf{x}_{l+r} = \mathbf{b}_{3l+2} + \mathbf{p}. \quad \blacksquare$$

2. INTERPOLATORY PERIODIC CUBIC B-SPLINE CURVES IN DE BOOR FORM

Because of (2) of page 11 $\mathbf{b}_{3l+2} = \dfrac{\Delta_{l+1}}{\Delta_l + \Delta_{l+1}}\mathbf{b}_{3l+1} + \dfrac{\Delta_l}{\Delta_l + \Delta_{l+1}}\mathbf{d}_l$ is true, resulting in the control points

$$\mathbf{d}_l = -\frac{\Delta_{l+1}}{\Delta_l}\mathbf{b}_{3l+1} + \frac{\Delta_l + \Delta_{l+1}}{\Delta_l}\mathbf{b}_{3l+2}.$$

DEFINITION 7.

$$d_{r,l,n} = -\frac{\Delta_{l+1}}{\Delta_l}h'_{r,l,n} + \frac{\Delta_l + \Delta_{l+1}}{\Delta_l}h''_{r,l,n}.$$

THEOREM 8. *For $d_{r,l,n}$, the following expression results by means of matrix multiplications*

$$d_{r,l,n} = \frac{\Delta_l\Delta_{l+1}}{2\Delta_{l+r}}\left(\frac{2}{\Delta_{l+r}}g^{\Delta_{l+r-1}}_{\Delta_l}(1,1) + g^{\Delta_{l+r-1}}_{\Delta_l}(2,1)\right) + \frac{\Delta_{l+1}-\Delta_l}{\Delta_{l+r}}\left(\frac{2}{\Delta_{l+r}}g^{\Delta_{l+r-1}}_{\Delta_l}(1,2) + g^{\Delta_{l+r-1}}_{\Delta_l}(2,2)\right) +$$

$$\frac{\Delta_l(\Delta_l + \Delta_{l+1})}{2\Delta_{l+r}}\left(\frac{2}{\Delta_{l+r}}g^{\Delta_{l+n-1}}_{\Delta_{l+r+1}}(2,2) + g^{\Delta_{l+n-1}}_{\Delta_{l+r+1}}(2,1)\right) + \frac{2\Delta_l + \Delta_{l+1}}{\Delta_{l+r}}\left(\frac{2}{\Delta_{l+r}}g^{\Delta_{l+n-1}}_{\Delta_{l+r+1}}(1,2) + g^{\Delta_{l+n-1}}_{\Delta_{l+r+1}}(1,1)\right).$$

Proof. The statement is obtained by using DEFINITION 6 for $h'_{r,l,n}$ and $h''_{r,l,n}$, by transforming $g^{\Delta_{l+r-1}}_{\Delta_l}(i,j)$ into an expression in terms of $g^{\Delta_{l+r-1}}_{\Delta_{l+1}}(i,j)$ (LEMMA 4), by transforming $g^{\Delta_{l+n}}_{\Delta_{l+r+1}}(i,j)$ into an expression in terms of $g^{\Delta_{l+n-1}}_{\Delta_{l+r+1}}(i,j)$, (LEMMA 3) and suitably combining. $\blacksquare$

COROLLARY 7. *For $d_{r,l,n}$ the explicit representation*

$$d_{r,l,n} = \delta_{0,r} + \frac{\Delta_l\Delta_{l+1}}{2\Delta_{l+r}}\left(-\frac{1}{\Delta_{l+r}}\sum_{i=0}^{\left[\frac{r-1}{2}\right]}3^i(-2)^{r-2i}K^{\Delta_{l+r-1}}_{\Delta_{l+1}}(2i) + \sum_{i=0}^{\left[\frac{r-2}{2}\right]}3^{i+1}(-2)^{r-2i-1}K_{\Delta_{l+1};\Delta_{l+r-1}}(2i+1)\right) +$$

$$\frac{\Delta_{l+1}-\Delta_l}{\Delta_{l+r}}\left(-\frac{1}{\Delta_{l+r}}\sum_{i=0}^{\left[\frac{r-2}{2}\right]}3^i(-2)^{r-2i-2}K^{\Delta_{l+1};\Delta_{l+r-1}}(2i+1) + \sum_{i=0}^{\left[\frac{r-1}{2}\right]}3^i(-2)^{r-2i-1}K^{\Delta_{l+1}}_{\Delta_{l+r-1}}(2i)\right) +$$

$$\frac{\Delta_l(\Delta_l + \Delta_{l+1})}{2\Delta_{l+r}}\left(-\frac{1}{\Delta_{l+r}}\sum_{i=0}^{\left[\frac{n-r-1}{2}\right]}3^i(-2)^{n-r-2i}K_{\Delta_{l+n-1}}^{\Delta_{l+r+1}}(2i)+\sum_{i=0}^{\left[\frac{n-r-2}{2}\right]}3^{i+1}(-2)^{n-r-2i-1}K_{\Delta_{l+r+1};\,l+n-1}(2i+1)\right)+$$

$$\frac{2\Delta_l + \Delta_{l+1}}{\Delta_{l+r}}\left(-\frac{1}{\Delta_{l+r}}\sum_{i=0}^{\left[\frac{n-r-2}{2}\right]}3^i(-2)^{n-r-2i-2}K^{\Delta_{l+r+1};\,\Delta_{l+n-1}}(2i+1)+\sum_{i=0}^{\left[\frac{n-r-1}{2}\right]}3^i(-2)^{n-r-2i-1}K_{\Delta_{l+r+1}}^{\Delta_{l+n-1}}(2i)\right)$$

is true.

COROLLARY 8. *For* $d_{r,l,n}$ *with* $\Delta_i = \Delta$ $(i = 0, 1, ..., n-1)$ *(equidistant parameters)*

$$d_{r,l,n} = \left(-\frac{1}{2}\right)^r\left((\sqrt{3}-1)^{2r-1}-(\sqrt{3}+1)^{2r-1}\right)-\left(-\frac{1}{2}\right)^{n-r+1}\left((\sqrt{3}-1)^{2n-2r+1}-(\sqrt{3}+1)^{2n-2r+1}\right)$$

is true. Thus $d_{r,l,n}$ *neither depends on* l *nor on* Δ_k.

Proof. Because of $d_{r,l,n} = -h'_{r,l,n} + 2h'_{r,l,n}$ (DEFINITION 7) and COROLLARY 6

$$d_{r,l,n} = \left(-\frac{1}{2}\right)^r(\sqrt{3}-1)^{2r-1}\left(-\frac{1}{6}\sqrt{3}(\sqrt{3}-1)^2-\frac{1}{3}\sqrt{3}\cdot(-2)\right)+$$

$$\left(-\frac{1}{2}\right)^r(\sqrt{3}+1)^{2r-1}\left(-\frac{1}{6}\sqrt{3}(\sqrt{3}+1)^2-\frac{1}{3}\sqrt{3}\cdot(-2)\right)+$$

$$\left(-\frac{1}{2}\right)^{n-r+1}(\sqrt{3}-1)^{2n-2r+1}\left(-\frac{1}{6}\sqrt{3}(-2)^2\frac{1}{(\sqrt{3}-1)^2}-\frac{1}{3}\sqrt{3}\cdot(-2)\right)+$$

$$\left(-\frac{1}{2}\right)^{n-r+1}(\sqrt{3}+1)^{2n-2r+1}\left(-\frac{1}{6}\sqrt{3}(-2)^2\frac{1}{(\sqrt{3}+1)^2}-\frac{1}{3}\sqrt{3}\cdot(-2)\right)$$

$$d_{r,l,n} = \left(-\frac{1}{2}\right)^r(\sqrt{3}-1)^{2r-1}-\left(-\frac{1}{2}\right)^r(\sqrt{3}+1)^{2r-1}+$$

$$\left(-\frac{1}{2}\right)^{n-r+1}(\sqrt{3}-1)^{2n-2r+1}\left(-\frac{2}{3}\sqrt{3}\left(\frac{-3+2\sqrt{3}}{4-2\sqrt{3}}\right)\right)+$$

$$\left(-\frac{1}{2}\right)^{n-r+1}(\sqrt{3}+1)^{2n-2r+1}\left(-\frac{2}{3}\sqrt{3}\left(\frac{-3-2\sqrt{3}}{4+2\sqrt{3}}\right)\right)$$

$$d_{r,l,n} = \left(-\frac{1}{2}\right)^r(\sqrt{3}-1)^{2r-1}-\left(-\frac{1}{2}\right)^r(\sqrt{3}+1)^{2r-1}+$$

$$\left(-\frac{1}{2}\right)^{n-r+1}(\sqrt{3}-1)^{2n-2r+1}\left(\frac{2\sqrt{3}-4}{4-2\sqrt{3}}\right)+\left(-\frac{1}{2}\right)^{n-r+1}(\sqrt{3}+1)^{2n-2r+1}\left(\frac{2\sqrt{3}+4}{4+2\sqrt{3}}\right). \quad\blacksquare$$

DEFINITION 8. *Recursive definition of the weight function* $N_{l,k}(t)$ *from de Boor (cf. Farin [1, 10.10 Rekursion von Mansfield, de Boor und Cox]):*

$$N_{l,0}(t) = \begin{cases}1 \text{ for } t_{l-1} \le t < t_l \\ 0 \text{ otherwise}\end{cases},$$

$$N_{l,k}(t) = \frac{(t - t_{l-1})N_{l,k-1}(t)}{t_{l+k-1} - t_{l-1}} + \frac{(t_{l+k} - t)N_{l+1,k-1}(t)}{t_{l+k} - t_l}.$$

THEOREM 9. *The interpolating periodic cubic spline s in* $[x_0; x_n]$ *can be represented as*

$$s = \sum_{l=-2}^{n} \mathbf{d}_l N_{l,3}(t).$$

(cf. Farin [1, (10.11)])
For t we have

$$... < t_{-2} < t_{-1} < t_0 < t_1 < t_2 < ...$$

and

$$t_{i+1} - t_i = \Delta_i.$$

For $\mathbf{d}_l$

$$\mathbf{d}_l = \mathbf{x}_l + \frac{1}{v_n}\sum_{r=0}^{n-1} d_{r,l,n}\Delta\mathbf{x}_{l+r}$$

is true.

3. NUMERICAL CALCULATION

With the formulas from THEOREM 7 and THEOREM 9, for great n, one quickly obtains great values in the nominators $h'_{r,l,n}$, $h''_{r,l,n}$, $d_{r,l,n}$ and in the denominator v_n. Therefore, the formulas from THEOREM 7 and THEOREM 9 are not very suitable for numerical calculation. This can be remedied using the following strategy. For this purpose, two definitions will be given plus LEMMA 10. In the following, let $c \in \mathbb{R}$ be a constant.

DEFINITION 9.

$$\overline{G}(a) = \begin{pmatrix} -\dfrac{2}{c} & -\dfrac{a}{2c} \\ -\dfrac{6}{ac} & -\dfrac{2}{c} \end{pmatrix} \quad ; \quad a > 0 \, ; c \in \mathbb{R} \ .$$

LEMMA 10. *The inverse of the matrix* $\overline{G}(a)$ *is*

$$\overline{G}^{-1}(a) = \begin{pmatrix} -2c & \dfrac{ac}{2} \\ \dfrac{6c}{a} & -2c \end{pmatrix} \quad ; \quad a > 0 \, ; c \in \mathbb{R} \ .$$

Proof.

$$\overline{G}(a) \cdot \overline{G}^{-1}(a) = I \quad ; \quad a > 0 \, ; c \in \mathbb{R} \ . \quad \blacksquare$$

DEFINITION 10. *If* $M = \{..., a_{-2}, a_{-1}, a_0, a_1, a_2, ...\}$ *is an ordered set of real positive numbers:*

(i)
$$\begin{pmatrix} \overline{g}_{a_i}^{a_j}(1,1) & \overline{g}_{a_i}^{a_j}(1,2) \\ \overline{g}_{a_i}^{a_j}(2,1) & \overline{g}_{a_i}^{a_j}(2,2) \end{pmatrix} = \overline{G}^{-1}(a_{j+1}) \cdot \overline{G}^{-1}(a_{j+2}) \cdot ... \cdot \overline{G}^{-1}(a_{i-1}) \ for \ i - j > 1 \ ,$$

(ii) $\qquad$ $\begin{pmatrix} \overline{g}_{a_i}^{a_j}(1,1) & \overline{g}_{a_i}^{a_j}(1,2) \\ \overline{g}_{a_i}^{a_j}(2,1) & \overline{g}_{a_i}^{a_j}(2,2) \end{pmatrix} = I \;\; for \;\; i-j=1 \;,$

(iii) $\qquad$ $\begin{pmatrix} \overline{g}_{a_i}^{a_j}(1,1) & \overline{g}_{a_i}^{a_j}(1,2) \\ \overline{g}_{a_i}^{a_j}(2,1) & \overline{g}_{a_i}^{a_j}(2,2) \end{pmatrix} = \overline{G}(a_j)\cdot\overline{G}(a_{j-1})\cdot...\cdot\overline{G}(a_i) \;\; for \;\; i-j<1 \;.$

Instead of $h'_{r,l,n}$, $h''_{r,l,n}$ and $d_{r,l,n}$ from THEOREM 7 and THEOREM 9, the following is used:

$$\overline{h}'_{r,l,n} = \frac{\Delta_l}{\Delta_{l+r}}\left(\frac{1}{c^{n-r}}\overline{g}_{\Delta_l}^{\Delta_{l+r-1}}(2,2)+\frac{1}{c^{r+1}}\overline{g}_{\Delta_{l+r+1}}^{\Delta_{l+n-1}}(1,1)\right)+\frac{2\Delta_l}{\Delta_{l+r}^2}\left(\frac{1}{c^{n-r}}\overline{g}_{\Delta_l}^{\Delta_{l+r-1}}(1,2)+\frac{1}{c^{r+1}}\overline{g}_{\Delta_{l+r+1}}^{\Delta_{l+n-1}}(1,2)\right)$$

$$\overline{h}''_{r,l,n} = -\frac{\Delta_l}{\Delta_{l+r}}\left(\frac{1}{c^{n-r+1}}\overline{g}_{\Delta_{l+1}}^{\Delta_{l+r-1}}(2,2)+\frac{1}{c^{r}}\overline{g}_{\Delta_{l+r+1}}^{\Delta_{l+n}}(1,1)\right)-\frac{2\Delta_l}{\Delta_{l+r}^2}\left(\frac{1}{c^{n-r+1}}\overline{g}_{\Delta_{l+1}}^{\Delta_{l+r-1}}(1,2)+\frac{1}{c^{r}}\overline{g}_{\Delta_{l+r+1}}^{\Delta_{l+n}}(1,2)\right)$$

and

$$\overline{d}_{r,l,n} = \frac{\Delta_l\Delta_{l+1}}{2\Delta_{l+r}c^{n-r+1}}\left(\frac{2}{\Delta_{l+r}}\overline{g}_{\Delta_{l+1}}^{\Delta_{l+r-1}}(1,1)+\overline{g}_{\Delta_{l+1}}^{\Delta_{l+r-1}}(2,1)\right)+\frac{\Delta_{l+1}-\Delta_l}{\Delta_{l+r}c^{n-r+1}}\left(\frac{2}{\Delta_{l+r}}\overline{g}_{\Delta_{l+1}}^{\Delta_{l+r-1}}(1,2)+\overline{g}_{\Delta_{l+1}}^{\Delta_{l+r-1}}(2,2)\right)+$$

$$\frac{\Delta_l(\Delta_l+\Delta_{l+1})}{2\Delta_{l+r}c^{r+1}}\left(\frac{2}{\Delta_{l+r}}\overline{g}_{\Delta_{l+r+1}}^{\Delta_{l+n-1}}(2,2)+\overline{g}_{\Delta_{l+r+1}}^{\Delta_{l+n-1}}(2,1)\right)+\frac{2\Delta_l+\Delta_{l+1}}{\Delta_{l+r}c^{r+1}}\left(\frac{2}{\Delta_{l+r}}\overline{g}_{\Delta_{l+r+1}}^{\Delta_{l+n-1}}(1,2)+\overline{g}_{\Delta_{l+r+1}}^{\Delta_{l+n-1}}(1,1)\right)$$

and instead of v_n from THEOREM 2

$$\overline{v}_n = \frac{2}{c^n}-\overline{g}_{\Delta_0}^{\Delta_{n-1}}(1,1)-\overline{g}_{\Delta_0}^{\Delta_{n-1}}(2,2).$$

This is equivalent to canceling the formulas from THEOREM 7 and THEOREM 9 by c^n. As c, the number

$$c = 2+\sqrt{3}$$

is suitable, because $\lim\limits_{n\to\infty}\dfrac{|v_n|}{(2+\sqrt{3})^n}=1$ for equidistant parameters. This will be shown in the following THEOREM 10.

THEOREM 10. With $\Delta_i = \Delta \;(i=0,1,...,n-1)$ (equidistant parameters)

$$c = \lim_{n\to\infty}\sqrt[n]{|v_n|} = 2+\sqrt{3}.$$

Proof. From COROLLARY 4 it can be seen that

$$|v_n| = (-1)^{n-1}\cdot 2+(2+\sqrt{3})^n+(2-\sqrt{3})^n.$$

Consequently:

$$\sqrt[n]{|v_n|} = \sqrt[n]{(-1)^{n-1}\cdot 2+(2+\sqrt{3})^n+(2-\sqrt{3})^n}.$$

It follows:

$$\lim_{n\to\infty}\sqrt[n]{\frac{|v_n|}{(2+\sqrt{3})^n}} = \lim_{n\to\infty}\sqrt[n]{-2\cdot\left(-\frac{1}{2+\sqrt{3}}\right)^n+1+\left(\frac{2-\sqrt{3}}{2+\sqrt{3}}\right)^n},$$

$$\lim_{n\to\infty}\frac{\sqrt[n]{|v_n|}}{2+\sqrt{3}} = 1$$

and with this the statement. ∎

This method can be accelerated by insertions of $\mathbf{d}_I$, which are calculated using the well-known iteration (cf. Farin [1, (9.6)])

$$\alpha_I \mathbf{d}_{I-1} + \beta_I \mathbf{d}_{I-2} + \gamma_I \mathbf{d}_I = (\Delta_{I-1} + \Delta_I)\mathbf{x}_I$$

with

$$\alpha_I = \frac{\Delta_I^2}{\Delta_{I-2} + \Delta_{I-1} + \Delta_I}$$

$$\beta_I = \frac{\Delta_I(\Delta_{I-2} + \Delta_{I-1})}{\Delta_{I-2} + \Delta_{I-1} + \Delta_I} + \frac{\Delta_{I-1}(\Delta_I + \Delta_{I+1})}{\Delta_{I-1} + \Delta_I + \Delta_{I+1}}$$

$$\gamma_I = \frac{\Delta_{I-1}^2}{\Delta_{I-1} + \Delta_I + \Delta_{I+1}}.$$

Not useful are insertions of more than 25 control points $\mathbf{d}_I$ by this method, because a great number of such insertions results in divergence.

The dependence of the computing time (in seconds) on the number of knots with 0, 10 and 20 values $\mathbf{d}_I$ inserted, which are generated using the above iteration, is shown the following diagram.

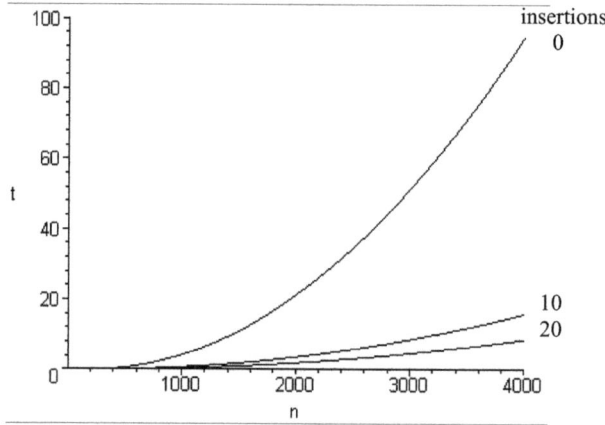

THEOREM 11. *After reducing* $\dfrac{h'_{r,l,n}}{v_n}$, $\dfrac{h''_{r,l,n}}{v_n}$ *and* $\dfrac{d_{r,l,n}}{v_n}$ *by* $(-1)^{n-1} \cdot (2+\sqrt{3})^n$ *with*

$\Delta_i = \Delta$ $(i = 0, 1, ..., n-1)$ *(equidistant parameters)*

$$\frac{h'_{r,l,n}}{v_n} = \frac{\sqrt{3}((-0.5)^r(\sqrt{3}-1)^{2r+1} + (-0.5)^{n-r-1}(\sqrt{3}-1)^{2n-2r-1})}{6 \cdot (1-(-0.5)^n(\sqrt{3}-1)^{2n})},$$

$$\frac{h''_{r,l,n}}{v_n} = -\frac{\sqrt{3} \cdot ((-0.5)^{r-1}(\sqrt{3}-1)^{2r-1} + (-0.5)^{n-r}(\sqrt{3}-1)^{2n-2r+1})}{6 \cdot (1-(-0.5)^n(\sqrt{3}-1)^{2n})}$$

and

$$\frac{d_{r,l,n}}{v_n} = \frac{(-0.5)^r(\sqrt{3}-1)^{2r-1} - (-0.5)^{n-r+1}(\sqrt{3}-1)^{2n-2r+1}}{1-(-0.5)^n(\sqrt{3}-1)^{2n}}$$

are true.

Proof. $\dfrac{1}{2+\sqrt{3}} = 2 - \sqrt{3} = \dfrac{1}{2}(\sqrt{3}-1)^2$. With this follows

$$\frac{1}{(-1)^{n-1} \cdot (2+\sqrt{3})^n} = (-1)^{-n+1} \cdot 0.5^n \cdot (\sqrt{3}-1)^{2n} = -(-0.5)^n \cdot (\sqrt{3}-1)^{2n}.$$

Because of COROLLARY 4 we get

$$\frac{v_n}{(-1)^{n-1} \cdot (2+\sqrt{3})^n} = (-0.5)^{n-1} \cdot (\sqrt{3}-1)^{2n} + 1 + 0.5^n \cdot (\sqrt{3}-1)^{2n}(2-\sqrt{3})^n.$$

As $2 - \sqrt{3} = \dfrac{1}{2}(\sqrt{3}-1)^2$

$$\frac{v_n}{(-1)^{n-1} \cdot (2+\sqrt{3})^n} = 1 + (-0.5)^{n-1}(\sqrt{3}-1)^{2n} + 0.5^{2n}(\sqrt{3}-1)^{4n}.$$

Written binomially

$$\frac{v_n}{(-1)^{n-1} \cdot (2+\sqrt{3})^n} = 1 - 2 \cdot (-0.5)^n(\sqrt{3}-1)^{2n} + (-0.5)^{2n}(\sqrt{3}-1)^{4n} = (1 - (-0.5)^n(\sqrt{3}-1)^{2n})^2.$$

Because of COROLLARY 6

$$h'_{r,l,n} = \frac{1}{6}\sqrt{3}\left(\left(-\frac{1}{2}\right)^r \left((\sqrt{3}+1)^{2r+1} + (\sqrt{3}-1)^{2r+1}\right) + \left(-\frac{1}{2}\right)^{n-r-1}\left((\sqrt{3}+1)^{2n-2r-1} + (\sqrt{3}-1)^{2n-2r-1}\right)\right).$$

In this formula, one can substitute $\sqrt{3}+1$ with $-(-0.5)^{-1}(\sqrt{3}-1)^{-1}$. This results in

$$\frac{h'_{r,l,n}}{(-1)^{n-1} \cdot (2+\sqrt{3})^n} = \frac{1}{6}\sqrt{3}\left((-0.5)^{n-r-1}(\sqrt{3}-1)^{2n-2r-1} - (-0.5)^{n+r}(\sqrt{3}-1)^{2n+2r+1} + \right.$$
$$\left. (-0.5)^r(\sqrt{3}-1)^{2r+1} - (-0.5)^{2n-r-1} \cdot (\sqrt{3}-1)^{4n-2r-1}\right).$$

Finally, the right side can be written as the product

$$\frac{h'_{r,l,n}}{(-1)^{n-1} \cdot (2+\sqrt{3})^n} = \frac{1}{6}\sqrt{3}(1 - (-0.5)^n(\sqrt{3}-1)^{2n}) \cdot ((-0.5)^r(\sqrt{3}-1)^{2r+1} + (-0.5)^{n-r-1}(\sqrt{3}-1)^{2n-2r-1})$$

Because of $h''_{r,l,n} = -h'_{r-1,l,n}$ (THEOREM 3 for equidistant parameters)

$$\frac{h''_{r,l,n}}{(-1)^{n-1} \cdot (2+\sqrt{3})^n} = -\frac{1}{6}\sqrt{3}(1 - (-0.5)^n(\sqrt{3}-1)^{2n}) \cdot ((-0.5)^{r-1}(\sqrt{3}-1)^{2r-1} + (-0.5)^{n-r}(\sqrt{3}-1)^{2n-2r+1})$$

is true.

Because of COROLLARY 8

$$d_{r,l,n} = \left(-\frac{1}{2}\right)^r \left((\sqrt{3}-1)^{2r-1} - (\sqrt{3}+1)^{2r-1}\right) - \left(-\frac{1}{2}\right)^{n-r+1}\left((\sqrt{3}-1)^{2n-2r+1} - (\sqrt{3}+1)^{2n-2r+1}\right).$$

The same substitution as mentioned above yields:

$$\frac{d_{r,l,n}}{(-1)^{n-1} \cdot (2+\sqrt{3})^n} = -(-0.5)^{n+r}(\sqrt{3}-1)^{2n+2r-1} - (-0.5)^{n-r+1}(\sqrt{3}-1)^{2n-2r+1} + \right.$$
$$(-0.5)^{2n-r+1}(\sqrt{3}-1)^{4n-2r+1} + (-0.5)^r(\sqrt{3}-1)^{2r-1}.$$

Finally, the right side can be written as the product:

$$\frac{d_{r,l,n}}{(-1)^{n-1} \cdot (2+\sqrt{3})^n} = (1 - (-0.5)^n(\sqrt{3}-1)^{2n}) \cdot ((-0.5)^r(\sqrt{3}-1)^{2r-1} - (-0.5)^{n-r+1}(\sqrt{3}-1)^{2n-2r+1}).$$

All expressions $\dfrac{v_n}{(-1)^{n-1}\cdot(2+\sqrt{3})^n}$, $\dfrac{h'_{r,l,n}}{(-1)^{n-1}\cdot(2+\sqrt{3})^n}$, $\dfrac{h''_{r,l,n}}{(-1)^{n-1}\cdot(2+\sqrt{3})^n}$ and

$\dfrac{d_{r,l,n}}{(-1)^{n-1}\cdot(2+\sqrt{3})^n}$ contain the factor $1-(-0.5)^n(\sqrt{3}-1)^{2n}$. Finally, the expression is reduced by this. ∎

COROLLARY 9 *With* $\Delta_i = \Delta$ $(i = 0, 1, ..., n-1)$ *(equidistant parameters)*

$$\lim_{n\to\infty}\frac{h'_{r,l,n}}{v_n} = \frac{\sqrt{3}(-0.5)^r(\sqrt{3}-1)^{2r+1}}{6},$$

$$\lim_{n\to\infty}\frac{h''_{r,l,n}}{v_n} = -\frac{\sqrt{3}(-0.5)^{r-1}(\sqrt{3}-1)^{2r-1}}{6}$$

and

$$\lim_{n\to\infty}\frac{d_{r,l,n}}{v_n} = (-0.5)^r(\sqrt{3}-1)^{2r-1}$$

are true.

4. REFERENCES

1. G. Farin, „Kurven und Flächen im Computer Aided Geometric Design Eine praktische Einführung," Vieweg Verlag, 1993. [In German]
2. F. Zeilfelder, „Interpolation und beste Approximation mit periodischen Splinefunktionen," Dissertation, Mannheim, 1996. [In German]

–